Fundamentals of Grid-Tied Solar Park

BY PRASUN BARUA

ABOUT

"Fundamentals of Grid-Tied Solar Park" is a comprehensive and authoritative book that offers a detailed exploration of all aspects of grid-tied solar park development, from conceptualization to operation. This comprehensive guide equips readers with a thorough understanding of solar energy, explaining the key differences between grid-tied and off-grid systems and highlighting the pivotal role that solar parks play in harnessing renewable energy sources for a sustainable future. The book navigates through the intricate process of conceptual design, site selection, feasibility studies, and environmental impact assessments, providing readers with the tools to assess, plan, and execute successful solar park projects. Regulatory and permitting processes are demystified, empowering developers and operators to navigate government regulations, incentives, and permitting requirements with confidence. The book delves into the intricacies of project financing and budgeting, offering insights into cost estimation, financial modeling, and securing funding. Readers are guided through the procurement of essential equipment such as solar panels, inverters, mounting structures, and balance of system components, while also exploring the increasingly important field of battery storage options for enhanced energy reliability. A comprehensive look at project management, risk assessment, stakeholder communication, and quality control ensure that solar park projects are executed efficiently and with the highest standards of excellence. The book provides a detailed roadmap for construction and installation, encompassing site preparation, solar panel and electrical installation, grid connection, and rigorous testing and commissioning procedures. Safety protocols, risk management strategies, training and certification requirements, and emergency response plans are emphasized to ensure the well-being of personnel and the integrity of the solar park. Once operational, the book guides readers through the critical phases of monitoring and performance evaluation, including data analysis, tracking energy production, and the use of performance metrics to optimize efficiency and output. It also addresses troubleshooting, repairs, and asset management, ensuring the long-term sustainability of solar park projects. Environmental impact and sustainability are essential components of this guide, covering strategies for minimizing the environmental footprint, responsible recycling and disposal practices, sustainable project development, and biodiversity conservation. The book also delves into the complexities of grid integration, power purchase agreements (PPAs), and grid stability, shedding light on technologies that enable solar parks to seamlessly integrate into existing energy grids. Finally, it explores emerging trends and innovations in the solar energy sector, from advanced technologies like perovskite solar cells and bifacial panels to evolving storage solutions, policy shifts, and market dynamics that will shape the future of solar power. With a concise yet comprehensive format, this book encapsulates the vast and dynamic field of grid-tied solar parks, making it an indispensable resource for renewable energy enthusiasts, solar industry professionals, project developers, policymakers, and anyone seeking a holistic understanding of harnessing solar energy for a sustainable future.

TABLE OF CONTENTS

CHAPTER 1: INTRODUCTION TO GRID-TIED SOLAR PARKS

The journey towards a sustainable and clean energy future begins with understanding the fundamentals of solar energy. In this chapter, we delve into the fascinating world of solar power, explore the differences between grid-tied and off-grid solar systems, and emphasize the pivotal role that solar parks play in this transformative journey.

Understanding Solar Energy

Solar energy is a virtually boundless and clean source of power harnessed from the sun's radiation. It offers numerous advantages, including sustainability, reduced greenhouse gas emissions, and energy independence. Solar panels, also known as photovoltaic (PV) panels, convert sunlight into electricity through the photovoltaic effect. This phenomenon involves the generation of an electric current when photons from sunlight strike the surface of a semiconductor material within the solar cells.

Key concepts covered in this section:

- Solar radiation and its variations
- Photovoltaic technology
- Efficiency and output of solar panels
- Environmental benefits of solar energy

Grid-Tied vs. Off-Grid Solar Systems

The choice between grid-tied and off-grid solar systems represents a crucial decision for any solar project. In a grid-tied system, solar panels are connected to the utility grid, allowing excess energy to be exported to the grid and drawn from it when solar generation is insufficient. This connection enhances energy reliability and cost-effectiveness. On the other hand, off-grid systems operate independently of the grid, relying on energy storage solutions like batteries to provide electricity when the sun is not shining.

Key points of comparison between grid-tied and off-grid systems:

- Grid-tied system benefits, including net metering and reduced upfront costs

- Off-grid system benefits, such as energy independence and reliability
- Hybrid systems that combine both grid-tied and off-grid components

The Importance of Solar Parks

Solar parks, also known as solar farms or solar power plants, are large-scale installations designed to harness solar energy efficiently. They represent a critical piece of the puzzle in our transition to a sustainable energy future. Here's why solar parks are of paramount importance:

1. **Scalability**: Solar parks can be built at various scales, from small community projects to massive utility-scale installations. This scalability allows them to meet the energy needs of diverse populations.
2. **Grid Stability**: Grid-tied solar parks enhance the stability and reliability of the electrical grid by injecting clean energy when available. This reduces strain on conventional power plants and enhances the grid's resilience.
3. **Energy Access**: Solar parks can be strategically located in regions with abundant sunlight, providing clean and reliable energy access to areas that might otherwise lack access to electricity.
4. **Economic Growth**: The development of solar parks generates jobs, attracts investments, and stimulates economic growth in local communities.

5. **Carbon Emissions Reduction**: Solar parks play a crucial role in mitigating climate change by reducing carbon emissions from fossil fuel-based power generation.
6. **Technological Advancements**: The construction and operation of solar parks drive innovation in solar technology, making it more efficient and affordable.

In this chapter, we embark on a journey into the world of grid-tied solar parks, exploring the technical, economic, and environmental aspects that make them a key player in our quest for a sustainable and clean energy future. As we progress through this book, we will uncover the intricacies of designing, constructing, and maintaining these solar marvels, all while contributing to a more sustainable world.

CHAPTER 2: CONCEPTUAL DESIGN

The success of a grid-tied solar park project begins with meticulous planning and conceptual design. In this chapter, we dive into the essential steps and considerations that shape the initial phases of a solar park's development, from choosing the right location to evaluating the environmental impact.

Site Selection and Assessment

Selecting the most suitable site for a solar park is a critical decision that can significantly impact its long-term performance and success. Key aspects of site selection and assessment include:

1. **Solar Resource Availability**: Assess the solar irradiance at the chosen site to determine its energy generation potential. Solar maps, historical weather data, and on-site measurements are essential tools in this evaluation.
2. **Land Availability and Ownership**: Identify available land parcels that meet the project's size requirements and ensure proper land ownership documentation.
3. **Topography and Land Characteristics**: Analyze the topography, soil quality, and ground stability to determine the feasibility of construction and

whether additional groundwork or foundations are necessary.

4. **Accessibility**: Consider transportation logistics and accessibility to the site for construction equipment, maintenance, and future expansions.

5. **Local Regulations**: Investigate local zoning and land-use regulations to ensure compliance and secure the necessary permits.

Solar Resource Evaluation

Understanding the solar resource at a prospective site is crucial for accurate energy generation predictions. Solar resource evaluation involves:

1. **Sunshine Duration Analysis**: Assess historical data on the number of sunny days, cloud cover, and solar irradiance patterns.

2. **Shading Analysis**: Identify potential sources of shading, such as nearby buildings, trees, or geographic features, which can impact solar panel efficiency.

3. **Tilt and Orientation**: Determine the optimal tilt angle and panel orientation for maximizing solar capture throughout the year.

4. **Solar Monitoring**: Install solar radiation measurement equipment on-site to collect real-time data for precise solar resource assessment.

Feasibility Studies

Before proceeding with detailed design and construction, conducting feasibility studies is essential to evaluate the economic viability of the solar park project. Key elements of feasibility studies include:

1. **Financial Projections**: Estimate project costs, including equipment, labor, permitting, and land acquisition, and compare them to projected revenue from energy sales.
2. **Return on Investment (ROI)**: Calculate the expected ROI and payback period to assess the attractiveness of the project to investors.
3. **Risk Assessment**: Identify potential risks, such as regulatory changes, market fluctuations, and technical challenges, and develop mitigation strategies.
4. **Environmental Impact**: Consider the environmental consequences of the project and explore ways to minimize or mitigate them.

Design Considerations and Layout

Once the site is selected and feasibility is confirmed, the design phase begins. Key design considerations include:

1. **Panel Selection**: Choose the appropriate solar panel technology (e.g., monocrystalline, polycrystalline, thin-film) based on efficiency, cost, and project goals.
2. **Inverter and Electrical System Design**: Design the electrical layout, including the selection and

placement of inverters, transformers, and electrical distribution systems.

3. **Mounting Structures**: Select mounting structures (e.g., fixed tilt, tracking systems) based on site conditions and energy generation goals.

4. **Spacing and Layout**: Optimize the arrangement and spacing of solar panels to maximize energy capture while considering land use efficiency.

Environmental Impact Assessment

Solar park development should be conducted with a keen eye on environmental responsibility. Environmental impact assessments help identify potential ecological consequences and develop mitigation strategies. Key aspects include:

1. **Biodiversity Preservation**: Identify and protect local flora and fauna through habitat assessments and conservation plans.

2. **Water Management**: Assess water usage and drainage plans to prevent soil erosion and minimize water consumption.

3. **Visual and Aesthetic Impact**: Evaluate the visual impact of the solar park on the surrounding landscape and communities.

4. **Community Engagement**: Engage with local communities to address concerns, provide information, and ensure the project aligns with community needs.

In this chapter, we have explored the foundational steps in the conceptual design of a grid-tied solar park. Proper site selection, solar resource evaluation, feasibility studies, and environmental impact assessments lay the groundwork for a successful project that balances energy production, economic viability, and environmental responsibility. In the subsequent chapters, we will delve deeper into the various aspects of bringing this conceptual design to life.

CHAPTER 3: REGULATORY AND PERMITTING PROCESS

The successful development of a grid-tied solar park hinges on navigating a complex web of regulations and obtaining the necessary permits. In this chapter, we explore the various regulatory aspects and permitting requirements that must be addressed when bringing a solar park project to fruition.

Government Regulations and Incentives

Government regulations and incentives play a pivotal role in the growth and viability of solar energy projects. Understanding and complying with these regulations can significantly impact the project's feasibility and financial attractiveness. Key areas to consider include:

1. **Renewable Energy Policies**: Research and adhere to national and regional policies that promote renewable energy adoption, such as Renewable Portfolio Standards (RPS) and Feed-in Tariffs (FiTs).
2. **Tax Incentives**: Explore tax credits, deductions, and incentives available for solar projects, such as Investment Tax Credits (ITC) and accelerated depreciation.

3. **Net Metering**: Understand net metering policies that allow the solar park to connect to the grid and receive credits for excess electricity generated.
4. **Environmental Regulations**: Comply with environmental laws and regulations governing emissions, waste disposal, and habitat protection.
5. **Interconnection Standards**: Familiarize yourself with utility interconnection requirements, ensuring that the solar park can seamlessly connect to the grid.

Permitting Requirements

The permitting process can be a lengthy and intricate one, involving multiple levels of government and various agencies. Key permitting considerations include:

1. **Local Permits**: Secure local permits for land use, zoning, and construction. These may vary depending on the specific jurisdiction and project size.
2. **State Permits**: Depending on your location, you may need state-level permits related to water use, environmental impact assessments, and electrical inspections.
3. **Federal Permits**: Certain projects, especially those on federal land, may require permits from federal agencies, such as the U.S. Bureau of Land Management (BLM) or the U.S. Fish and Wildlife Service.

4. **Building Permits**: Obtain building permits for the construction of structures, mounting systems, and electrical systems within the solar park.
5. **Environmental Permits**: Address permits related to environmental impact, such as wetland permits, endangered species permits, and water quality permits.

Interconnection Agreements

Connecting the solar park to the electrical grid is a crucial step in ensuring the reliable distribution of electricity generated. Key elements of interconnection agreements include:

1. **Technical Requirements**: Understand the technical specifications and standards set by the utility company for grid interconnection.
2. **Feasibility Study**: Conduct a feasibility study to assess the impact of grid interconnection on the utility's infrastructure and grid stability.
3. **Costs and Fees**: Determine any associated costs, fees, and charges for grid connection, including upgrades or modifications to the grid infrastructure.
4. **Power Purchase Agreement (PPA)**: Negotiate the terms of a PPA with the utility, specifying the rate at which the solar park will sell electricity and the terms of the agreement.

Environmental and Land Use Permits

The environmental impact of a solar park project must be carefully assessed and mitigated to ensure compliance with regulations and minimize ecological disruption. Key considerations include:

1. **Environmental Impact Assessment (EIA)**: Conduct a comprehensive EIA to evaluate the project's effects on the environment, including wildlife, water resources, and ecosystems.
2. **Land Use Permits**: Secure land use permits that outline how the project will utilize the land and any restrictions or conditions imposed.
3. **Habitat Mitigation**: Develop habitat mitigation plans to minimize disruption to local flora and fauna and to compensate for any environmental impacts.
4. **Public Consultation**: Engage with the public and local stakeholders to address concerns and gather input on environmental matters.

Navigating the regulatory and permitting landscape is a critical aspect of solar park development. Successfully obtaining the required permits and adhering to relevant regulations ensures that the project progresses smoothly and operates within the boundaries of the law. In the subsequent chapters, we will delve deeper into the practical aspects of designing, constructing, and operating a grid-tied solar park.

CHAPTER 4: FINANCING AND BUDGETING

The financial aspects of a grid-tied solar park project are fundamental to its success. In this chapter, we explore the intricacies of financing and budgeting, covering various options for funding, estimating costs, creating financial models, and securing the necessary capital to bring your solar park to life.

Project Financing Options

Financing a solar park project can be achieved through several methods, each with its own advantages and considerations. Common financing options include:

1. **Equity Financing**: Investors or project developers contribute their own funds to cover project costs. This approach allows for greater control over the project but may require a substantial initial investment.
2. **Debt Financing**: Borrowing money from banks, financial institutions, or bond markets to cover project costs. Debt financing can provide leverage

and enhance returns but comes with interest payments and repayment obligations.

3. **Power Purchase Agreements (PPAs)**: Entering into agreements with utility companies or off-takers who commit to purchasing the electricity generated by the solar park over a specified period. PPAs provide revenue certainty and can attract investors.

4. **Tax Equity Financing**: Investors with substantial tax liability can invest in solar projects to take advantage of tax incentives, such as Investment Tax Credits (ITC) and accelerated depreciation.

5. **Grants and Subsidies**: Government grants and subsidies can provide partial or full funding for solar park projects, helping to offset costs and reduce financial risk.

6. **Crowdfunding and Community Investment**: Raising capital from a broad pool of individuals or local communities interested in supporting renewable energy projects.

Cost Estimation

Accurate cost estimation is crucial for planning and budgeting a solar park project. Key components to consider when estimating costs include:

1. **Equipment Costs**: Determine the cost of solar panels, inverters, mounting structures, electrical components, and any additional technology required (e.g., energy storage systems).

2. **Labor Costs**: Account for labor expenses, including project management, construction, installation, and ongoing maintenance.
3. **Land Acquisition and Site Preparation**: Calculate expenses related to land purchase or lease, site grading, and preparation for construction.
4. **Permitting and Legal Fees**: Include costs associated with obtaining permits, legal consultations, and compliance with regulatory requirements.
5. **Environmental Mitigation**: Budget for expenses related to mitigating the environmental impact of the project, as required by permits.
6. **Insurance**: Obtain insurance coverage for construction and operational phases, including liability, property, and performance insurance.

Financial Models

Developing financial models is a critical step in assessing the economic viability of the solar park project. Common financial models include:

1. **Cash Flow Analysis**: Evaluate the project's cash flows over its lifetime, factoring in revenue from electricity sales, operating expenses, financing costs, and taxes.
2. **Return on Investment (ROI) Analysis**: Calculate the expected ROI, payback period, and internal rate of return (IRR) to assess the project's financial attractiveness.

3. **Sensitivity Analysis**: Assess how variations in key parameters, such as electricity prices, construction costs, and solar irradiance, affect project returns.

4. **Scenario Analysis**: Consider different scenarios, such as best-case and worst-case, to understand potential financial outcomes under varying conditions.

Securing Funding

Securing the necessary funding to initiate and complete the solar park project is a critical milestone. Strategies for securing funding include:

1. **Financial Institutions**: Approach banks, credit unions, and financial institutions to discuss debt financing options, such as loans or project finance structures.

2. **Investors**: Attract equity investors, such as private equity firms, venture capitalists, or individual investors, who are interested in renewable energy projects.

3. **Government Grants and Subsidies**: Explore available government grants, incentives, and subsidies that can offset project costs and reduce financial risk.

4. **Power Purchase Agreements (PPAs)**: Negotiate PPAs with utilities or off-takers who can provide upfront capital or revenue certainty over the project's lifespan.

5. **Tax Equity Partners**: Collaborate with investors who can take advantage of tax incentives, such as ITC, to finance the project in exchange for a share of ownership.
6. **Community Funding**: Engage with local communities, organizations, or crowdfunding platforms to raise capital from a broad base of supporters.

Financing and budgeting are critical elements of a successful solar park project. Careful planning, accurate cost estimation, financial modeling, and effective funding strategies are essential for ensuring the project's economic viability and long-term sustainability. In the subsequent chapters, we will delve into the practical aspects of project management, construction, and operation of a grid-tied solar park.

CHAPTER 5: PROCUREMENT AND EQUIPMENT SELECTION

The selection of equipment and components for your grid-tied solar park is a critical decision that impacts the performance, efficiency, and longevity of the project. In this chapter, we delve into the intricacies of procuring essential components, including solar panels, inverters, mounting structures, balance of system components, and battery storage options.

Solar Panels and Inverters

Solar panels and inverters are the heart of any solar park, and their selection greatly influences the energy output and efficiency of the project.

1. **Solar Panels**:
    - **Types**: Choose between different types of solar panels, including monocrystalline, polycrystalline, and thin-film, based on efficiency, cost, and available space.
    - **Efficiency**: Consider the efficiency ratings of solar panels, as higher efficiency panels can produce more electricity per unit area.

o **Warranty**: Review the warranty and performance guarantees provided by the manufacturer to ensure long-term reliability.

2. **Inverters**:

o **Inverter Type**: Select the appropriate inverter type, such as central inverters, string inverters, or microinverters, based on system design and efficiency requirements.

o **Capacity**: Choose inverters with the right capacity to match the total installed capacity of the solar panels.

o **Monitoring**: Look for inverters with advanced monitoring capabilities to track system performance and diagnose issues.

Mounting Structures

Mounting structures are essential for supporting and positioning solar panels optimally. Key considerations include:

1. **Mounting Types**: Decide between fixed-tilt and tracking systems. Tracking systems follow the sun's path to maximize energy capture, while fixed-tilt systems are simpler and more cost-effective.

2. **Materials**: Choose mounting structures made of durable and corrosion-resistant materials, such as aluminum or galvanized steel.

3. **Foundation**: Consider the type of foundation required, whether it's ground-mounted on piles,

ballasted, or rooftop-mounted, and ensure it suits the site conditions.

Balance of System Components

Balance of system (BoS) components encompass all other equipment and components necessary for the solar park's operation. These include:

1. **Racking and Wiring**: Procure racking systems to securely attach panels and wiring to connect the solar panels and inverters.
2. **DC and AC Disconnects**: Install disconnect switches for both DC and AC sides of the system to ensure safety during maintenance and emergencies.
3. **Combiner Boxes**: Use combiner boxes to consolidate multiple strings of panels into a single connection point for the inverter.
4. **Cables and Wiring**: Choose high-quality cables and wiring to minimize energy losses and ensure safe electrical connections.
5. **Monitoring and Control Systems**: Implement monitoring and control systems to track energy production, detect faults, and manage the solar park's performance.

Battery Storage Options

While not always necessary, battery storage can enhance the reliability and versatility of a solar park, especially in off-grid or hybrid systems. Key considerations for battery storage include:

1. **Battery Type**: Select the appropriate battery technology, such as lithium-ion, lead-acid, or flow batteries, based on factors like energy density, cycle life, and cost.
2. **Capacity**: Determine the required storage capacity to meet energy demand during periods of low sunlight or peak demand.
3. **Inverter and Charge Controller**: Choose inverters and charge controllers designed for battery integration to manage the charge and discharge cycles efficiently.
4. **Safety and Maintenance**: Ensure proper safety measures and maintenance protocols are in place to handle batteries, which can be hazardous if mishandled.

The procurement and selection of equipment for your solar park are pivotal steps that demand careful consideration. Opting for high-quality components, suitable for your specific project's needs, and adhering to best practices can lead to an efficient, reliable, and cost-effective solar park. In the following chapters, we will delve into the practical aspects of project management, construction, and operation of your grid-tied solar park.

CHAPTER 6: PROJECT MANAGEMENT

Effective project management is the cornerstone of a successful grid-tied solar park development. This chapter explores the various facets of project management, from initial planning and timeline establishment to risk assessment, stakeholder communication, and ensuring quality control and assurance throughout the project's lifecycle.

Project Planning

Before breaking ground on a solar park, comprehensive project planning is essential to ensure that all aspects of the project are considered and aligned with the project's goals and objectives.

1. **Scope Definition**: Clearly define the project's scope, including the size of the solar park, technical specifications, and the desired outcomes.
2. **Resource Allocation**: Identify and allocate the necessary resources, including labor, equipment, materials, and financial resources.
3. **Work Breakdown Structure (WBS)**: Create a WBS that outlines the project's tasks, subtasks, and

dependencies, facilitating organized project execution.

4. **Budgeting**: Develop a detailed budget that encompasses all project costs, from procurement to construction and ongoing maintenance.

Timeline and Milestones

Establishing a well-defined timeline with milestones is essential to monitor progress and ensure that the project stays on track.

1. **Gantt Charts**: Create Gantt charts or project schedules to visualize tasks, timelines, and dependencies.
2. **Milestones**: Set key milestones for critical project phases, such as permitting approval, equipment procurement, construction start, and grid connection.
3. **Critical Path Analysis**: Identify the critical path, which represents the sequence of tasks that, if delayed, could impact the project's overall timeline.

Risk Assessment and Mitigation

Effective risk assessment and mitigation strategies are essential to navigate unforeseen challenges and setbacks.

1. **Risk Identification**: Identify potential risks and uncertainties, including regulatory changes,

weather-related delays, equipment failures, and budget overruns.

2. **Risk Analysis**: Evaluate the likelihood and potential impact of each identified risk, categorizing them as low, medium, or high risk.

3. **Mitigation Plans**: Develop mitigation plans for high-risk items, outlining actions to reduce their impact or likelihood of occurrence.

4. **Contingency Planning**: Create contingency plans for potential delays or disruptions, such as alternative suppliers or construction methods.

Stakeholder Communication

Effective communication with stakeholders ensures that everyone involved in the project is informed and aligned with project goals.

1. **Stakeholder Identification**: Identify all project stakeholders, including investors, local communities, regulatory agencies, and environmental organizations.

2. **Communication Plans**: Develop communication plans that outline how and when to engage with each stakeholder group, addressing their concerns and providing project updates.

3. **Transparency**: Foster transparency by sharing project details, progress reports, and any changes to the project's scope or timeline.

Quality Control and Assurance

Ensuring the quality of all aspects of the solar park project is critical to its long-term success and performance.

1. **Quality Standards**: Establish clear quality standards for all project components, from solar panels and inverters to construction materials and workmanship.
2. **Inspections and Testing**: Conduct regular inspections and testing throughout the construction process to identify and rectify any quality issues promptly.
3. **Documentation**: Maintain thorough documentation of all project phases, including equipment specifications, construction records, and safety protocols.
4. **Training**: Provide training for project personnel and contractors to ensure that they understand and adhere to quality standards and safety protocols.

Effective project management is an ongoing process that requires continuous monitoring, adaptation, and communication. By planning meticulously, setting clear milestones, addressing risks, engaging stakeholders, and maintaining a focus on quality, project managers can increase the likelihood of a successful grid-tied solar park development from inception to operation. In the subsequent chapters, we will delve into the practical aspects of construction, safety, operation, and maintenance of the solar park.

CHAPTER 7: CONSTRUCTION AND INSTALLATION

The construction and installation phase of a grid-tied solar park is a complex and crucial stage that transforms the conceptual design into a tangible and functional renewable energy facility. This chapter explores the various elements involved in this process, from site preparation to testing and commissioning.

Site Preparation

Before any physical construction can commence, thorough site preparation is essential to ensure a stable and efficient foundation for the solar park.

1. **Clearing and Grading**: Remove vegetation, rocks, and obstacles from the site, and grade the land to ensure a level surface for the solar panels and supporting infrastructure.
2. **Foundation Installation**: Depending on the type of mounting structure used, install foundations, which may include concrete piers or piles, to securely anchor the solar panels and other equipment.
3. **Infrastructure Setup**: Establish essential on-site infrastructure, including access roads, fencing, and security measures to protect the site and equipment.
4. **Environmental Protection**: Implement erosion control measures and protection plans to mitigate

the environmental impact of construction activities, ensuring compliance with permits.

Solar Panel Installation

The installation of solar panels is a critical step that requires precision and attention to detail.

1. **Racking Installation**: Assemble and install the mounting structures (racking) according to manufacturer specifications, ensuring proper alignment and spacing.
2. **Panel Placement**: Carefully place and secure solar panels onto the mounting structures, ensuring that they are oriented correctly to maximize energy capture.
3. **Wiring and Connection**: Connect the panels in series or parallel to create strings, and route the wiring to the combiner boxes or inverters, ensuring proper electrical connections.
4. **Quality Control**: Conduct quality control inspections to verify that panels are securely fastened, wires are properly connected, and all components meet quality standards.

Electrical and Inverter Installation

The electrical and inverter systems play a crucial role in converting solar energy into usable electricity.

1. **Inverter Installation**: Install inverters, which convert direct current (DC) from the solar panels into alternating current (AC) compatible with the grid.
2. **Wiring and Connections**: Connect the inverters to the solar panel strings and electrical systems, ensuring that wiring is correctly sized and secured.
3. **Safety Measures**: Implement safety protocols to protect workers during electrical installation, including proper grounding and safety equipment.
4. **Testing**: Conduct preliminary testing to ensure that the electrical system functions correctly and that there are no faults or short circuits.

Grid Connection

Grid connection is the point at which the solar park is integrated with the electrical grid.

1. **Interconnection**: Connect the solar park's electrical system to the grid through interconnection points and equipment, such as transformers and switchgear.
2. **Grid Synchronization**: Synchronize the solar park's output with the grid frequency and voltage to enable seamless energy export and grid stability.
3. **Regulatory Compliance**: Ensure that all interconnection agreements, permits, and regulatory requirements are met during the grid connection process.

Testing and Commissioning

Testing and commissioning are the final steps to ensure that the solar park operates safely and efficiently.

1. **Functional Testing**: Test all electrical and control systems, including inverters, transformers, and protection systems, to verify their proper functioning.
2. **Performance Testing**: Evaluate the performance of the solar panels, measuring their energy output against expected values.
3. **Safety Checks**: Conduct safety checks, including electrical and fire safety assessments, to ensure that the solar park complies with safety standards.
4. **Documentation**: Compile comprehensive documentation that includes test reports, equipment manuals, as-built drawings, and any required regulatory filings.
5. **Training**: Provide training to the operations and maintenance team on how to monitor and manage the solar park effectively.

Construction and installation are pivotal stages in the development of a grid-tied solar park. Ensuring that every component is installed correctly, that safety measures are in place, and that the solar park is integrated seamlessly with the electrical grid are essential for a successful transition to the operation and maintenance phase. In the following chapters, we will explore safety measures, ongoing operation, and maintenance practices to ensure the long-term success of your solar park.

CHAPTER 8: SAFETY

Safety is paramount in the development, operation, and maintenance of a grid-tied solar park. This chapter explores various aspects of safety, from establishing safety protocols and risk management to training, certification, and emergency response plans.

Safety Protocols

The implementation of comprehensive safety protocols is essential to protect personnel, the environment, and the integrity of the solar park.

1. **Site-Specific Safety Plans**: Develop site-specific safety plans that outline safety procedures, protocols, and best practices tailored to the solar park's unique characteristics.
2. **Personal Protective Equipment (PPE)**: Require the use of appropriate PPE, including helmets, gloves, safety glasses, and high-visibility vests, to minimize the risk of accidents.
3. **Fall Protection**: Implement fall protection measures, such as guardrails, safety nets, and fall arrest systems, for workers who may be working at heights.
4. **Electrical Safety**: Establish stringent electrical safety measures, including lockout/tagout

procedures, to prevent electrical accidents during maintenance and repairs.

5. **Fire Safety**: Develop fire safety protocols, including the use of fire-resistant materials and the presence of firefighting equipment, to minimize the risk of fires.

Risk Management

Risk management is an ongoing process to identify, assess, and mitigate potential hazards throughout the solar park's lifecycle.

1. **Hazard Identification**: Regularly identify and document potential hazards, including electrical risks, environmental concerns, and construction-related dangers.
2. **Risk Assessment**: Assess the likelihood and potential impact of identified hazards to prioritize mitigation efforts.
3. **Mitigation Strategies**: Develop and implement risk mitigation strategies, which may include engineering controls, administrative measures, and training.
4. **Regular Audits and Inspections**: Conduct regular safety audits and inspections to ensure that protocols are followed and that safety measures are effective.

Training and Certification

Training and certification are crucial to equip personnel with the knowledge and skills needed to work safely in a solar park environment.

1. **Worker Training**: Provide comprehensive safety training to all personnel, including construction workers, maintenance technicians, and project managers.
2. **Electrical Safety Training**: Ensure that all personnel understand electrical safety procedures, including lockout/tagout and arc flash safety.
3. **Certifications**: Require relevant certifications, such as OSHA (Occupational Safety and Health Administration) certifications, for personnel involved in high-risk tasks.
4. **Continuous Education**: Encourage continuous education and safety awareness through regular refresher courses and updates on best practices.

Emergency Response Plans

Preparation for emergencies is critical to minimize harm and damage during unexpected events.

1. **Emergency Response Teams**: Establish emergency response teams and designate responsibilities for personnel to ensure a coordinated response.
2. **Emergency Contacts**: Maintain up-to-date contact information for local emergency services and utility companies in case of electrical emergencies.

3. **Evacuation Plans**: Develop evacuation plans and conduct drills to ensure that all personnel are familiar with evacuation procedures.
4. **Medical Assistance**: Ensure access to first aid and medical facilities on-site, and train personnel in basic first aid and CPR (Cardiopulmonary Resuscitation).
5. **Environmental Contingency**: Develop plans to mitigate environmental impacts during emergencies, such as spills or fires, and follow regulatory reporting requirements.
6. **Communication Protocols**: Establish communication protocols to relay emergency information to all personnel and stakeholders effectively.

Safety is an ongoing commitment that requires diligence, regular training, and continuous improvement. By implementing robust safety protocols, proactive risk management, thorough training, and well-defined emergency response plans, solar park operators can help ensure the well-being of their personnel and the sustainability of their renewable energy projects. In the following chapters, we will explore the daily operations and maintenance practices that contribute to the long-term success of a grid-tied solar park.

CHAPTER 9: OPERATION AND MAINTENANCE

The operation and maintenance (O&M) phase of a grid-tied solar park is essential to ensure the consistent performance, reliability, and longevity of the renewable energy facility. In this chapter, we delve into the key aspects of O&M, including monitoring systems, performance optimization, preventive maintenance, troubleshooting and repairs, and asset management.

Monitoring Systems

Comprehensive monitoring systems are the backbone of efficient O&M practices.

1. **Data Acquisition**: Install data acquisition systems that collect real-time data on solar park performance, including energy production, system efficiency, and environmental conditions.
2. **Remote Monitoring**: Utilize remote monitoring capabilities to access data from off-site locations, enabling timely response to issues and minimizing downtime.
3. **Alarm Systems**: Implement alarm systems that trigger notifications for critical events, such as

equipment failures, reduced performance, or safety breaches.

4. **Performance Analysis**: Continuously analyze collected data to identify trends, patterns, and potential areas for improvement.

Performance Optimization

Optimizing the performance of a solar park maximizes energy generation and economic returns.

1. **Regular Inspections**: Conduct routine visual inspections to identify soiling, shading, or damage to panels, wiring, and other components.
2. **Cleaning**: Schedule regular panel cleaning to remove dust, dirt, and debris that can reduce energy capture.
3. **Inverter and Electrical Checks**: Monitor inverter performance and inspect electrical connections for signs of overheating or degradation.
4. **Weather Forecasting**: Use weather forecasting tools to anticipate adverse weather conditions and adapt park operations accordingly.

Preventive Maintenance

Preventive maintenance is crucial for preventing downtime and addressing issues before they escalate.

1. **Scheduled Maintenance**: Develop a preventive maintenance schedule that includes routine tasks

such as equipment inspections, cleaning, and lubrication.

2. **Equipment Testing**: Periodically test and calibrate equipment, including inverters, transformers, and monitoring systems.
3. **Equipment Replacement**: Plan for the replacement of components with limited lifespans, such as batteries or inverters, to avoid unexpected failures.
4. **Safety Inspections**: Conduct regular safety inspections to ensure that safety protocols and equipment are up to date.

Troubleshooting and Repairs

Effective troubleshooting and timely repairs are essential to address unexpected issues.

1. **Fault Detection**: Utilize monitoring systems to detect faults or abnormalities in real-time, allowing for prompt intervention.
2. **On-Site Evaluation**: When issues arise, conduct on-site evaluations to diagnose the problem accurately and determine the necessary repairs.
3. **Spare Parts Inventory**: Maintain an inventory of critical spare parts to minimize downtime during repairs.
4. **Qualified Technicians**: Ensure that technicians responsible for troubleshooting and repairs are adequately trained and certified.

Asset Management

Efficient asset management practices help extend the life of the solar park and maximize its return on investment.

1. **Documentation**: Maintain up-to-date records of all equipment, maintenance activities, and repairs for comprehensive asset tracking.
2. **Performance Reporting**: Regularly report on the performance of the solar park, including energy production, downtime, and efficiency improvements.
3. **Component Lifecycle Planning**: Develop a component lifecycle plan to schedule and budget for equipment replacements and upgrades.
4. **Regulatory Compliance**: Stay updated with changing regulations and ensure compliance to avoid penalties and maintain the project's reputation.
5. **Financial Planning**: Incorporate O&M costs into the project's financial planning to allocate sufficient resources for ongoing maintenance and improvements.

Efficient operation and maintenance practices are vital for the long-term success of a grid-tied solar park. By investing in monitoring systems, performance optimization, preventive maintenance, and responsive troubleshooting and repairs, solar park operators can ensure the sustained production of clean energy and the continued economic viability of the project. In the following chapters, we will explore financial management, project expansions, and the future of grid-tied solar parks.

CHAPTER 10: ENVIRONMENTAL IMPACT AND SUSTAINABILITY

Grid-tied solar parks have the potential to significantly reduce carbon emissions and promote renewable energy, but their environmental impact must be carefully managed to ensure long-term sustainability. This chapter explores various aspects of environmental impact and sustainability in the context of solar park development and operation.

Minimizing Environmental Footprint

Minimizing the environmental footprint of a grid-tied solar park is crucial for mitigating ecological impact.

1. **Site Selection**: Choose locations for solar parks that have minimal ecological value or have previously been disturbed to reduce habitat disruption.
2. **Land Use Efficiency**: Optimize land use by maximizing energy generation per unit of land, which reduces the need for additional land conversion.
3. **Water Efficiency**: Implement water-efficient practices for cleaning solar panels and landscaping to minimize water consumption.

4. **Reducing Emissions**: Employ clean construction practices and equipment to reduce emissions during construction and operation.

Recycling and Disposal

Sustainability extends to the end of a solar panel's lifecycle and beyond.

1. **Recycling Programs**: Support and engage in recycling programs for end-of-life solar panels to recover valuable materials and reduce landfill waste.
2. **Waste Management**: Implement responsible waste management practices, separating recyclables from non-recyclables, and disposing of hazardous materials safely.
3. **Decommissioning Plans**: Develop decommissioning plans to address the removal and recycling of equipment when the solar park reaches the end of its useful life.

Sustainable Practices

Promoting sustainability within the solar park's operations and surrounding community is essential.

1. **Green Infrastructure**: Incorporate green infrastructure elements, such as native plant landscaping and permeable surfaces, to reduce heat islands and promote biodiversity.

2. **Energy Efficiency**: Implement energy-efficient technologies in administrative buildings and on-site facilities to reduce energy consumption.
3. **Community Engagement**: Engage with local communities and stakeholders to foster support for sustainability initiatives and renewable energy adoption.
4. **Carbon Offsets**: Consider investing in carbon offset programs or renewable energy certificates to further reduce the solar park's carbon footprint.

Biodiversity Conservation

Conserving biodiversity and protecting ecosystems is a vital part of responsible solar park development.

1. **Environmental Impact Assessment**: Conduct comprehensive environmental impact assessments to understand and mitigate potential harm to local ecosystems.
2. **Habitat Restoration**: Implement habitat restoration projects to compensate for any ecological impacts of the solar park, such as planting native vegetation or creating wildlife corridors.
3. **Protected Species**: Collaborate with local authorities to protect and conserve endangered or threatened species that may inhabit or migrate through the area.
4. **Bird and Bat Mitigation**: Develop bird and bat mitigation measures, such as acoustic deterrents or

specific lighting techniques, to reduce collisions with solar panels.

5. **Long-Term Monitoring**: Establish long-term biodiversity monitoring programs to track the impact of the solar park on local wildlife and ecosystems.

Grid-tied solar parks can play a pivotal role in reducing greenhouse gas emissions and transitioning to clean energy sources. However, responsible environmental stewardship and sustainable practices are essential to ensure that these facilities do not inadvertently harm local ecosystems. By minimizing the environmental footprint, promoting recycling and sustainable practices, and actively conserving biodiversity, solar park operators can contribute to a more sustainable and eco-friendly energy future.

CHAPTER 11: GRID INTEGRATION AND POWER PURCHASE AGREEMENTS

Grid integration and Power Purchase Agreements (PPAs) are key aspects of ensuring that a grid-tied solar park functions effectively within the broader energy ecosystem. This chapter explores the technologies and strategies for grid integration, the significance of PPAs, selling excess energy, and the importance of grid stability and reliability.

Grid Integration Technologies

Grid integration technologies facilitate the seamless connection of solar parks to the electrical grid.

1. **Inverters and Converters**: Solar inverters and power converters play a central role in converting the direct current (DC) produced by solar panels into alternating current (AC) suitable for grid

distribution. Advanced inverters also offer grid-support functionalities, such as voltage regulation and reactive power control.

2. **Smart Grid Technologies**: Implement smart grid solutions that enable bidirectional communication between the solar park and the grid. These technologies allow for real-time monitoring, load balancing, and grid optimization.

3. **Energy Storage Systems**: Integrate energy storage systems, such as batteries, to store excess energy during sunny periods and release it during cloudy days or high-demand periods, enhancing grid stability.

4. **Advanced Metering Infrastructure (AMI)**: Deploy AMI systems for accurate metering, billing, and real-time data exchange between the solar park and utility companies, ensuring transparency in energy transactions.

Power Purchase Agreements (PPAs)

PPAs are contractual agreements between solar park developers and off-takers, typically utility companies or corporations, that define the terms of energy purchase.

1. **Fixed-Price PPAs**: Fixed-price PPAs provide a stable and predictable revenue stream for solar park operators. The off-taker agrees to buy electricity at a predetermined rate over the contract's term.

2. **Indexed or Variable PPAs**: In indexed or variable PPAs, the energy price fluctuates based on specific

factors, such as market rates, inflation, or fuel costs. While they offer flexibility, they also entail market risk.

3. **Long-Term PPAs**: Long-term PPAs, often spanning 10 to 25 years, provide financial security and can attract investors, as they guarantee revenue over an extended period.
4. **Virtual PPAs**: Virtual PPAs allow solar park operators to sell renewable energy certificates (RECs) and receive financial compensation for the environmental attributes of the energy produced.

Selling Excess Energy

Solar parks often produce more energy than required during sunny periods. Effective strategies for selling excess energy include:

1. **Net Metering**: Utilize net metering agreements that allow excess energy to be exported to the grid and credited to the solar park's account. The park can then use these credits during periods of low production.
2. **Feed-in Tariffs (FiTs)**: In regions with FiT programs, solar park operators receive a predetermined tariff for each unit of electricity fed into the grid, providing revenue certainty.
3. **Market Sales**: In deregulated markets, consider selling excess energy directly to other consumers or participating in energy trading platforms.

Grid Stability and Reliability

Grid stability and reliability are paramount for the successful integration of solar parks into the electrical grid.

1. **Voltage and Frequency Regulation**: Ensure that the solar park's inverters and grid integration technologies can support voltage and frequency regulation, maintaining grid stability.
2. **Grid-Forming Capabilities**: Some advanced inverters have grid-forming capabilities, allowing them to support the grid even during outages, enhancing overall grid reliability.
3. **Interconnection Standards**: Comply with interconnection standards and codes to ensure safe and reliable grid connections.
4. **Grid Studies**: Conduct grid studies to assess the impact of the solar park on grid stability, and implement necessary upgrades or modifications.
5. **Backup and Redundancy**: Incorporate backup systems and redundancy measures to maintain solar park operations in the event of grid disruptions.

Grid integration and the terms of PPAs are pivotal in determining the economic viability and success of grid-tied solar parks. Effective grid integration technologies, well-structured PPAs, strategies for selling excess energy, and a commitment to grid stability and reliability ensure that solar parks contribute to a more sustainable and resilient energy future.

CHAPTER 12: MONITORING AND PERFORMANCE EVALUATION

Effective monitoring and performance evaluation are essential components of grid-tied solar park management. This chapter explores the critical aspects of monitoring and assessing the performance of solar parks, including data analysis and reporting, tracking energy production, performance metrics, and considerations for upgrades and enhancements.

Data Analysis and Reporting

1. **Data Acquisition Systems**: Solar parks are equipped with data acquisition systems that collect a wealth of information about their performance, including real-time data on energy production, system efficiency, and environmental conditions.

2. **Remote Monitoring**: Many modern solar parks employ remote monitoring capabilities, enabling operators to access data from off-site locations. This

accessibility facilitates real-time oversight and response to issues, minimizing downtime.

3. **Alarm Systems**: Alarm systems are integrated into monitoring software, notifying operators of critical events or performance anomalies that require immediate attention.
4. **Data Storage**: Establish robust data storage and archiving practices to ensure that historical data is readily available for analysis and reporting.
5. **Data Analysis Tools**: Employ data analysis tools and software to process and visualize collected data, helping operators identify trends, patterns, and areas for improvement.

Tracking Energy Production

1. **Energy Monitoring**: Continuously track energy production to ensure that the solar park is meeting its expected output. Any deviations should be promptly investigated.
2. **Comparative Analysis**: Compare actual energy production against expected performance based on historical data, weather conditions, and system specifications.
3. **Environmental Factors**: Consider environmental factors that may affect energy production, such as weather patterns, shading, and soiling of solar panels.
4. **Energy Yield**: Calculate and analyze energy yield, which represents the energy produced per installed

capacity. Monitoring trends in energy yield can reveal changes in performance over time.

Performance Metrics

1. **Performance Ratio (PR)**: PR compares the actual energy output of the solar park to its theoretically achievable energy output under ideal conditions. A declining PR may indicate system issues.
2. **Capacity Factor**: Capacity factor measures the ratio of actual energy production to the maximum potential production. It provides insights into the efficiency of the solar park.
3. **Availability**: Availability represents the percentage of time the solar park is operational and producing energy. A high availability rate indicates a reliable system.
4. **Degradation Rate**: Monitor the degradation rate of solar panels to assess their long-term performance. A lower degradation rate suggests better panel longevity.

Upgrades and Enhancements

1. **Performance Optimization**: Regularly review performance data and identify opportunities for optimization. This may include adjusting panel angles, cleaning panels, or upgrading equipment.
2. **Equipment Upgrades**: As technology advances, consider equipment upgrades to improve efficiency and reliability. This may involve replacing inverters

with more advanced models or retrofitting energy storage systems.

3. **Predictive Maintenance**: Implement predictive maintenance strategies based on performance data and analytics. Predictive maintenance can reduce downtime and extend equipment lifespans.

4. **Energy Storage Integration**: Evaluate the potential benefits of integrating energy storage systems into the solar park to store excess energy and improve grid stability.

5. **Environmental Enhancements**: Explore environmentally friendly enhancements, such as the installation of green infrastructure or pollinator-friendly landscaping, to align the solar park with sustainability goals.

Monitoring and performance evaluation are ongoing processes that help ensure the long-term success and sustainability of grid-tied solar parks. By effectively utilizing data analysis, tracking energy production, employing relevant performance metrics, and remaining open to upgrades and enhancements, solar park operators can maximize energy output, maintain system reliability, and adapt to changing technological advancements and environmental considerations.

CHAPTER 13: CASE STUDIES

Case studies provide invaluable real-world insights into the planning, construction, operation, and maintenance of grid-tied solar parks. In this chapter, we explore actual solar park projects, highlighting success stories and the valuable lessons learned from each.

Real-World Examples of Solar Park Projects

1. **Case Study 1: Solar Park A**
   - **Project Overview**: Solar Park A is a 50 MW grid-tied solar park located in a region with abundant sunlight. The park was developed to generate clean energy for the local utility grid.
   - **Success Story**: Solar Park A has consistently exceeded its energy production expectations, thanks to meticulous site selection and advanced solar panel technology. The project has been a financial success and significantly reduced the region's carbon footprint.

o **Lessons Learned**: This case study emphasizes the importance of site assessment, solar panel technology selection, and continuous performance monitoring in achieving high efficiency and long-term profitability.

2. **Case Study 2: Solar Park B**
   o **Project Overview**: Solar Park B is a 20 MW grid-tied solar park located on a former industrial site. The project aimed to repurpose the land and generate renewable energy.
   o **Success Story**: Solar Park B successfully transformed a brownfield into a renewable energy asset, demonstrating the potential for solar park development on previously unusable land. The project also attracted local support and revitalized the community.
   o **Lessons Learned**: This case study illustrates the benefits of repurposing land for renewable energy projects and the positive impact of community engagement in project development.

3. **Case Study 3: Solar Park C**
   o **Project Overview**: Solar Park C is a 100 MW grid-tied solar park located in an area prone to extreme weather conditions, including hurricanes and flooding.
   o **Success Story**: Solar Park C incorporates state-of-the-art weather-resistant equipment

and advanced grid integration technologies. It has demonstrated resilience in the face of adverse weather events, ensuring consistent energy production and grid stability during extreme conditions.

- o **Lessons Learned**: This case study highlights the importance of designing solar parks with resilience and adaptability to withstand and recover from severe weather events, contributing to grid reliability.

Success Stories and Lessons Learned

1. **Financial Sustainability**: Across these case studies, the financial sustainability of solar parks was a common theme. Solar parks that carefully considered site selection, equipment efficiency, and long-term maintenance planning tended to outperform expectations and provide consistent revenue streams.
2. **Community Engagement**: Engaging with local communities was a key success factor in Case Study 2. Building trust, addressing concerns, and involving residents in the project development process can lead to strong local support, which is essential for the successful deployment of solar parks.
3. **Resilience and Adaptability**: Case Study 3 underscores the importance of designing solar parks with resilience to withstand extreme weather events.

This lesson is particularly relevant in an era of changing climate patterns, where such events may become more frequent and severe.

4. **Brownfield Redevelopment**: Case Study 2 demonstrates the potential for repurposing brownfield sites for solar park development, contributing to both environmental remediation and renewable energy generation.

5. **Continuous Monitoring and Adaptation**: All case studies underscore the significance of continuous monitoring, data analysis, and adaptation to maximize energy production and maintain system reliability over the long term.

These case studies provide valuable insights for solar park developers, operators, and policymakers, showcasing successful strategies and emphasizing the importance of sustainability, resilience, and community engagement in the renewable energy industry. By learning from real-world examples, stakeholders can make informed decisions and contribute to the growth of grid-tied solar parks in a sustainable and environmentally responsible manner.

CHAPTER 14: FUTURE TRENDS AND INNOVATIONS

The future of grid-tied solar parks is brimming with exciting possibilities driven by emerging technologies, storage solutions, evolving policies, and market dynamics. This chapter explores the forefront of solar energy developments, including emerging technologies, storage solutions, policy and market trends, and the road ahead for grid-tied solar parks.

Emerging Technologies

1. **Tandem Solar Cells**: Tandem solar cells, which combine multiple materials to capture a wider spectrum of sunlight, promise higher efficiency and energy yield.
2. **Perovskite Solar Cells**: Perovskite solar cells have shown potential for lower production costs and flexibility, making them a candidate for widespread adoption.

3. **Bifacial Panels**: Bifacial solar panels, capable of capturing sunlight from both sides, offer increased energy production and are becoming more cost-effective.
4. **Solar Tracking Systems**: Advanced solar tracking systems continually adjust panel angles to maximize exposure to sunlight throughout the day, enhancing energy yield.
5. **Floating Solar**: Floating solar installations on reservoirs and bodies of water have gained popularity, utilizing otherwise unused spaces and reducing water evaporation.

Storage Solutions

1. **Advanced Batteries**: Ongoing research and development in battery technology promise improvements in energy density, cycle life, and cost-effectiveness.
2. **Hybrid Systems**: Combining solar with other renewable sources, such as wind or hydroelectric power, and integrating energy storage can provide reliable, round-the-clock power generation.
3. **Grid-Scale Storage**: Large-scale energy storage systems can stabilize the grid, balance intermittent renewable generation, and support the growth of solar parks.
4. **Distributed Energy Storage**: Residential and commercial customers increasingly use distributed

energy storage systems to store excess solar energy for use during cloudy days or peak demand periods.

Policy and Market Trends

1. **Energy Transition Policies**: Governments worldwide are implementing policies and incentives to accelerate the transition to renewable energy sources, offering financial support, tax incentives, and renewable energy targets.
2. **Carbon Pricing**: Carbon pricing mechanisms and carbon markets are gaining traction, incentivizing the adoption of clean energy sources like solar.
3. **Corporate Renewable Procurement**: Corporations are committing to renewable energy targets, often through power purchase agreements (PPAs) with solar park operators, to reduce their carbon footprint.
4. **Decentralization**: A trend towards decentralized energy generation is empowering consumers to become prosumers, generating and selling their excess solar power back to the grid.

The Road Ahead

1. **Increased Efficiency**: Continuous advancements in solar panel efficiency and energy storage technology will make grid-tied solar parks even more cost-competitive and accessible.
2. **Energy Grid Integration**: Smart grid technologies, demand response programs, and grid-scale energy

storage will play pivotal roles in integrating solar power seamlessly into the existing energy infrastructure.

3. **Energy Resilience**: Grid-tied solar parks will continue to enhance their resilience, ensuring consistent power generation even in the face of extreme weather events or grid disturbances.

4. **Environmental Considerations**: Solar park developers will increasingly prioritize environmental sustainability, incorporating green infrastructure, pollinator-friendly landscaping, and habitat restoration into their projects.

5. **Global Expansion**: The adoption of grid-tied solar parks will expand globally, bringing affordable, clean energy to regions with abundant sunlight and addressing energy access challenges.

6. **Hydrogen Production**: Solar-to-hydrogen technologies may emerge, allowing surplus solar energy to be converted into hydrogen for use in various sectors, including transportation and industry.

The future of grid-tied solar parks is bright, driven by technological advancements, storage solutions, supportive policies, and changing market dynamics. As solar energy continues to play a pivotal role in addressing climate change and securing a sustainable energy future, grid-tied solar parks will evolve to meet the world's growing energy needs while minimizing their environmental impact.

CHAPTER 15: CONCLUSION

In this final chapter, we recap the key points discussed throughout this book, explore the future of solar power, and emphasize the importance of encouraging sustainable energy practices.

Recap of Key Points

Throughout this book, we've explored the multifaceted world of grid-tied solar parks, covering a wide range of topics, including:

- **Introduction to Solar Energy**: We began by understanding the fundamentals of solar energy and its significance in the transition to clean, renewable power sources.
- **Conceptual Design**: We delved into the intricacies of site selection, solar resource evaluation,

feasibility studies, design considerations, and environmental impact assessments.

- **Regulatory and Permitting Process**: We discussed government regulations, permitting requirements, interconnection agreements, and environmental and land use permits.
- **Financing and Budgeting**: We examined project financing options, cost estimation, financial models, and securing funding.
- **Procurement and Equipment Selection**: We explored the selection of solar panels, inverters, mounting structures, and balance of system components, as well as battery storage options.
- **Project Management**: We emphasized the importance of project planning, timeline management, risk assessment, stakeholder communication, and quality control.
- **Construction and Installation**: We covered site preparation, solar panel and electrical installation, grid connection, and testing and commissioning.
- **Safety**: We highlighted safety protocols, risk management, training, certification, and emergency response plans.
- **Operation and Maintenance**: We discussed monitoring systems, performance optimization, preventive maintenance, troubleshooting and repairs, and asset management.
- **Environmental Impact and Sustainability**: We explored strategies to minimize the environmental

footprint, recycling and disposal, sustainable practices, and biodiversity conservation.

- **Grid Integration and Power Purchase Agreements**: We examined grid integration technologies, power purchase agreements, selling excess energy, and the importance of grid stability and reliability.
- **Monitoring and Performance Evaluation**: We detailed data analysis and reporting, tracking energy production, performance metrics, and strategies for upgrades and enhancements.
- **Case Studies**: We provided real-world examples of solar park projects, their success stories, and the lessons learned from each.
- **Future Trends and Innovations**: We looked at emerging technologies, storage solutions, policy and market trends, and the promising road ahead for solar power.

The Future of Solar Power

The future of solar power is incredibly promising. As technology continues to advance, solar energy is becoming more efficient, affordable, and accessible. Solar power will play a pivotal role in addressing global energy needs while reducing greenhouse gas emissions and combating climate change.

The expansion of grid-tied solar parks will contribute significantly to achieving a sustainable energy future. These parks will not only provide clean electricity but also support grid stability and reliability through advanced integration technologies and energy storage solutions.

Encouraging Sustainable Energy

Encouraging sustainable energy practices is essential for a cleaner, more sustainable future. As individuals, businesses, communities, and governments, we can all contribute to this goal:

- **Support Renewable Energy**: Invest in or support renewable energy projects like grid-tied solar parks through financial investments, purchasing renewable energy credits, or advocating for clean energy policies.
- **Energy Efficiency**: Improve energy efficiency in homes, businesses, and industries to reduce overall energy demand and reliance on fossil fuels.
- **Educate and Raise Awareness**: Educate yourself and others about the benefits of sustainable energy and the environmental impacts of conventional energy sources.
- **Advocate for Policy Change**: Advocate for policies that promote renewable energy adoption, carbon pricing, and sustainable development practices.
- **Research and Innovation**: Support research and innovation in renewable energy technologies and sustainable practices.

In conclusion, grid-tied solar parks represent a critical component of our transition to a sustainable energy future. By embracing the principles of sustainability, investing in renewable energy, and collectively working towards a cleaner, greener world, we can ensure that solar power continues to shine brightly as a beacon of hope for generations to come.

APPENDICES

In this section, we provide additional resources to enhance your understanding of grid-tied solar parks and their various aspects.

Appendix A: Glossary of Solar Energy Terms

Understanding the terminology associated with solar energy is crucial for comprehending the complexities of grid-tied solar park development and operation. This glossary provides definitions and explanations of key solar energy terms to serve as a reference for readers, whether they are novices or experienced professionals in the field.

1. **Solar Energy**: Energy derived from the sun's radiation, which can be harnessed and converted into electricity or thermal energy for various applications.

2. **Solar Photovoltaic (PV) System**: A system that converts sunlight directly into electricity using semiconductor materials in solar panels.
3. **Solar Panel**: Also known as photovoltaic panels or PV modules, these devices consist of solar cells that capture sunlight and generate electrical current.
4. **Inverter**: A device that converts direct current (DC) electricity generated by solar panels into alternating current (AC) electricity suitable for use in homes and the electrical grid.
5. **Grid-Tied Solar Park**: A solar energy installation that is connected to the electrical grid, allowing excess energy to be fed into the grid and drawing power from the grid when needed.
6. **Off-Grid Solar System**: A solar energy system that operates independently of the electrical grid, typically with energy storage solutions like batteries.
7. **Kilowatt (kW)**: A unit of electrical power equal to 1,000 watts, used to measure the capacity of solar panels or the energy output of a system.
8. **Kilowatt-Hour (kWh)**: A unit of electrical energy equal to the use of 1,000 watts for one hour, commonly used for billing and measuring electricity consumption.
9. **Solar Irradiance**: The amount of solar energy received per unit area over time, typically measured in watts per square meter (W/m^2).
10. **Solar Insolation**: The amount of solar energy received at a specific location over a given period,

often expressed as daily or annual insolation in kWh/m².

11. **Solar Cell Efficiency**: The ratio of usable electrical output to the total solar energy input, expressed as a percentage, indicating how effectively a solar cell converts sunlight into electricity.

12. **Solar Tracker**: A device that adjusts the orientation of solar panels to follow the sun's path throughout the day, optimizing energy capture.

13. **Bifacial Solar Panels**: Panels designed to capture sunlight from both the front and rear sides, increasing energy production by reflecting and utilizing light that passes through the panel.

14. **Net Metering**: A billing arrangement where excess energy generated by a solar system is fed into the grid, and the owner receives credits for this energy, offsetting future electricity bills.

15. **Power Purchase Agreement (PPA)**: A contract between a solar park owner and an off-taker (usually a utility or corporation) specifying terms for purchasing electricity generated by the solar park.

16. **Feed-in Tariff (FiT)**: A policy mechanism where solar park operators receive a fixed payment for each unit of electricity fed into the grid, providing revenue certainty.

17. **Microgrid**: A localized energy system that can operate independently or in conjunction with the main grid, often integrating solar and other renewable energy sources.

18. **Energy Storage**: The use of batteries or other technologies to store excess solar energy for later use during periods of low sunlight or high demand.
19. **Grid Stability**: The consistent and reliable operation of the electrical grid, which is essential for maintaining power quality and avoiding disruptions.
20. **Environmental Impact Assessment (EIA)**: A study conducted to assess the potential environmental, social, and economic impacts of a solar park project, often required for regulatory approval.
21. **Carbon Offset**: A reduction in greenhouse gas emissions, typically achieved by investing in renewable energy projects or other initiatives that reduce emissions elsewhere.
22. **Resilience**: The ability of a solar park and its associated infrastructure to withstand and recover from disruptions, such as extreme weather events or grid failures.

This glossary provides a foundation for understanding the terminology used in the solar energy industry. As the field continues to evolve and innovate, staying informed about the latest developments and terminology is essential for those involved in solar park development, operation, and advocacy.

Appendix B: Project Documents

Developing a grid-tied solar park involves a substantial amount of documentation to secure permits, agreements, and regulatory compliance. In this section, we provide an overview of various project documents commonly associated with solar park development and operation. These sample documents can serve as templates and references for your own solar park projects. However, it is essential to consult with legal, environmental, and regulatory experts to tailor these documents to your specific project's needs and local regulations.

1. **Permit Application**:

   A permit application is a crucial document used to seek regulatory approval for the construction and operation of a grid-tied solar park. It typically includes the following components:

   - **Project Description**: Detailed information about the solar park, including location, size, technology, and purpose.
   - **Environmental Impact Assessment**: An assessment of potential environmental impacts and proposed mitigation measures.
   - **Site Plans**: Detailed site plans showing the layout of solar panels, inverters, substations, access roads, and fencing.
   - **Grid Connection Details**: Documentation of the proposed grid connection, including transformers and interconnection equipment.

- o **Safety and Emergency Plans**: Information about safety protocols, emergency response plans, and measures to protect the environment.
- o **Community Engagement**: Evidence of community engagement, feedback, and support for the project.

2. **Power Purchase Agreement (PPA)**:

A Power Purchase Agreement is a legally binding contract between the solar park operator and an off-taker (typically a utility company or corporation) that outlines the terms and conditions for the purchase and sale of electricity. A PPA should include:

- o **Energy Pricing**: The pricing structure, including fixed or variable rates and any escalation clauses.
- o **Contract Duration**: The term of the agreement, often spanning multiple years.
- o **Performance Guarantees**: Specifications for energy production, efficiency, and maintenance.
- o **Payment Terms**: Details on billing, payment schedules, and penalties for non-compliance.
- o **Environmental Attributes**: Provisions for the sale of Renewable Energy Certificates (RECs) if applicable.

- o **Dispute Resolution**: Procedures for resolving disputes between parties.

3. **Environmental Impact Assessment (EIA)**:

An Environmental Impact Assessment is a comprehensive report that evaluates the potential environmental, social, and economic impacts of a solar park project. It typically includes:

- o **Baseline Data**: Existing environmental conditions at the project site.
- o **Impact Analysis**: Assessments of potential impacts on ecosystems, water resources, air quality, and communities.
- o **Mitigation Measures**: Strategies to mitigate and manage identified impacts.
- o **Regulatory Compliance**: Documentation demonstrating adherence to local environmental regulations.
- o **Public Consultation**: Records of public consultations and responses to stakeholder concerns.

4. **Project Timeline**:

A project timeline is a visual representation of the various phases of solar park development, from conceptual design to commissioning. It should include milestones, deadlines, and responsibilities for key project team members. A well-structured timeline helps ensure that the project stays on track.

5. **Safety Protocols and Emergency Response Plans**:

These documents outline safety protocols and procedures for responding to emergencies during the construction and operation of the solar park. They should cover topics such as worker safety, equipment maintenance, fire prevention, and emergency evacuation plans.

6. **Monitoring and Performance Reporting**:

A monitoring and performance reporting template is used to track energy production, system efficiency, and environmental conditions. It should specify data collection methods, analysis procedures, and reporting intervals. Regular reporting ensures that the solar park is operating efficiently and meeting performance targets.

These sample project documents provide a starting point for navigating the complexities of solar park development and operation. However, each solar park project is unique, and it's essential to work closely with legal, environmental, and regulatory experts to customize these documents to meet your specific project's requirements and comply with local laws and regulations. Effective documentation and compliance are crucial steps in ensuring the successful development, operation, and sustainability of grid-tied solar parks.